THE CASE
FOR LEGIBILITY

THE CASE
FOR LEGIBILITY

JOHN RYDER

THE BODLEY HEAD
LONDON   SYDNEY
TORONTO

British Library Cataloguing
in Publication Data
Ryder, John, b. 1917
The case for legibility
1. Legibility (Printing)
2. Book design
I. Title
686.2'24   Z250.A4

ISBN 0-370-30158-7
© John Ryder 1979
Printed in Great Britain for
The Bodley Head Ltd
9 Bow Street, London WC2E 7AL
by The Stellar Press, Hatfield
Set in 'Monotype' Ehrhardt
First published 1979

AUTHOR'S NOTE

When asked by the Bodleian Library in 1976 if I would speak at the Standing Conference of National and University Libraries, I agreed with reluctance. From that moment until the autumn of 1977, when the talk had to be delivered in the Clarendon Building, Oxford, I lived in fear and trembling of what I had consented to do. But by the spring of the following year I had twice more given my talk in the City of Oxford. While working on the revisions for my paper on legibility, The Bodley Head asked me if it could be adapted for issue in book form. The present text with its monochrome illustrations is my answer.

In the making of books the agent of communication between the author and the reader is the designer, and from the appearance of printed books it is not always clear that designers understand and practise what is required of them in order to establish this communication. For me this is 'the case for legibility', which implies not just a readable page but a complete book produced in such a way that it is easy to use and to read.

With all the new developments in the printing industry (like computer-assisted filmsetting and cathode-ray tube font development) we need designers wedded to the pursuit of legibility, for it must never be taken for granted that legibility is universally desired. To some it is detestable. New alphabets are drawn in ways which make them difficult to read. Advertisements are designed and printed to the bewilderment of all who notice them and then avert their eyes without receiving the message. In the process of book design legibility becomes

(1) *Michael Harvey's drawing of a normal e surrounded by variants some of which are unpleasant, low-legibility characters*

important as soon as the author's work is taken seriously, or as soon as the effect of design upon the eye of the reader is considered.

Since 1935 I have devoted much time to written and printed images on paper, and so have come to know something about the process of book design and about typographical matters in general. Recent startling changes in techniques have not invalidated the canons of good typographical arrangement, though they have made sympathetic understanding between designer and printer more difficult to establish and maintain.

Once legibility is accepted as the objective, the urge to achieve it may be expressed in many ways: for instance in the act of making the eye of an e open enough to be read (1). Such careful attention to detail must be applied to every written and printed letter, and not only to the letter itself but also to how and where it appears on the page – in relation to adjacent letters, words, lines of letters, margins.

The process of design for legibility can be described in three stages:

1. knowing and understanding the author's and editor's intention,
2. translating this intention into typographical signs and instructions so that setting and proofing may be done,
3. arranging the typographical material on proof into a sequence of pages which, to the best of the designer's skill and experience, reflects the author's intention.

This process begins with the choice of an alphabet for the text setting. At the same time format and paper must be considered. Then the text is arranged into typographical material which will make a sequence of pages, taking account of illustration, decoration and space wherever applicable. Now a specification and schedule of work and materials is made and this involves 'finishing' – not just details of binding, but the whole process of imposition and printing as well.

Throughout this procedure the designer can only achieve coherence and unity by subjecting every detail to the test of suitability – suitability to subject, author, publisher, market, mechanical processes of origination and reproduction, and, of course, to costing

– closely controlled costs are part of each stage of design.

These statements imply freedom of choice but in practice the designer's choice is restricted. For many books, the economics of publishing have already chosen the typesetter. The choice of typeface is therefore narrowed. Then speed of production may exclude some faces whose matrices or keybars or discs are not immediately available. If we try to be positive and choose Baskerville, we soon find a need to be more specific, for that simple type name may refer to:

Monotype Series 169 metal letters

or Monophoto film images

or Lino or Intertype metal slugs

or Linotype VIP film

or Linotron CRT scanned lines

or Autologic (APS4) CRT scanned dots

or IBM golfball electronic typewriting.

A complex of considerations will make the choice. Moreover, the closest attention will have to be paid to proofing, platemaking and inking controls if the designer's intentions are not to be frustrated, even by presses of good reputation.

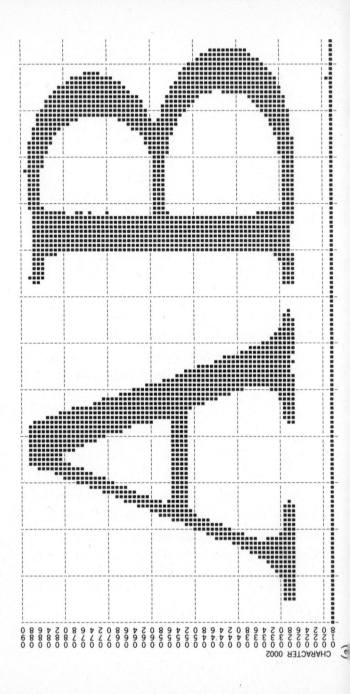

If you cannot choose and control the apparent typeface, how can you choose a suitable paper for it?

Perhaps the imposition of the type area on the page, the margins, can be exactly given? No. These details are not constant because of the variation in machine folding of large sheets, and in any case often not precise because of the economic necessity of producing sewn and perfect-bound copies from one and the same imposition.

Type size sounds like a detail within the designer's control but it lacks precision and objectivity when we have to talk in terms of linear percentages of an original which may itself be an image projected to film by a light source; or dots or lines scanned electronically, projected photographically (2); or a proof from hot metal setting reduced or enlarged by camera. Only print from type carries the certainty of measurable type size.

Face, size, impression, paper, imposition – is there normally such lack of precision as to blur the designer's intention? I think there is. Therefore if we are striving for legibility we must get back to first principles

(2) *An unedited dot map produced by scanning and showing the need for additions and deletions*

and accept the control that these principles impose.

The prognosis is not good. Discipline is needed to save the eyesight of the human race. If that sounds like overstatement, just reflect that the adoption of dry transfer lettering (3) and CRT typesetting have given the designer scope beyond his knowledge and experience, for lettering design is not taught in most colleges of art, and this sad state of affairs will continue as long as attention to the roman alphabet is thought unnecessary once we are out of the nursery.

When my connections with typography began in 1935, I had left school and begun work in a London bookshop. At this time I bought a typewriter and started to make books in typescript; bought a handpress and a miniature font of Gill Sans, and began to print. Then in 1940 I was reluctantly joined to the Army and remained for a long time as much bewildered by typographical points as by the violence of political volcanoes erupting in Europe. Enforced travels abroad put an end to all studies except that I learnt to milk a parachute and to assist surgeons.

Appeal

illiterate

HOCKEY

STICK

SLAVISH

PUNCH-DRUNK

FAVOUR

OGRE

(3) *Dry transfer lettering misused*

The theatre of war was sometimes terrifying; but the present terror of being overcome by illiteracy and blinded by illegibility is just as real. Before the war my knowledge of typography was too slight for me to know what I was really doing. I now know that whatever is to be done must be done for the reader's benefit and for maximum legibility.

So the heart of the matter is that if writing is to be read, then problems of legibility must be solved. You must present an author's work to the reader without fuss and with design techniques as invisible as possible. This concept surely requires no defence.

A page in a book must not look like a collection of letters. At least it should look like a collection of words, at best a collection of phrases. To achieve this, not only must the right letter-design for the language and text and format be chosen, but it must also be used in a way which allows the easy flow, and therefore recognition, of phrases. That is the basis of designing a page for the printer to print for the reader to read.

The image shown here (4) of a Latin text

tius in eadem miseria vivant tardiusque moriantur. Procul dubio ergo indicant, inmortalitatem, saltem talem quae non habeat finem mendicitatis, quanta gratulatione susciperent. Quid? animalia omnia etiam inrationalia, quibus datum non est ista cogitare, ab inmensis draconibus usque ad exiguos vermiculos nonne se esse velle atque ob hoc interitum fugere omnibus quibus possunt motibus indicant? Quid? arbusta omnesque frutices, quibus nullus est sensus ad vitandam manifesta motione perniciem, nonne ut in auras tutum cacuminis germen emittant, aliud terrae radicis adfigunt, quo alimentum trahant atque ita suum quodam modo esse conservent? Ipsa postremo corpora, quibus non solum sensus, sed nec ulla saltem seminalis est vita, ita tamen vel exiliunt in superna vel in ima descendunt vel librantur in mediis, ut essentiam suam, ubi secundum naturam possunt esse, custodiant.

Iam vero nosse quantum ametur quamque falli nolit humana natura, vel hinc intellegi potest, quod lamentari quisque sana mente mavult quam laetari in amentia. Quae vis magna atque mirabilis mortalibus praeter homini animantibus nulla est, licet eorum quibusdam ad istam lucem contuendam multo quam nobis sit acrior sensus oculorum; sed lucem illam incorpoream contingere nequeunt, qua mens nostra quodam modo radiatur, ut de his omnibus recte iudicare possimus. Nam in quantum eam capimus, in tantum id possumus. Verum tamen inest in sensibus inrationalium animantium, etsi scientia nullo modo, at certe quaedam scientiae similitudo; cetera autem rerum corporalium, non

(4) *Caslon types set and printed by the Chiswick Press in 1912*

qui omnibus ui aquarum submersis cum fi
mirabili quodā modo quasi semen huāni ge
utinā quasi uiuam quandam imaginem imi
quidem ante diluuium fuerunt:post diluuiu
altissimi dei sacerdos iustitiæ ac pietatis mira
bræorū appellatus est:apud quos nec circun
ulla mentio erat . Quare nec iudæos(posteris
gentiles:quoniam non ut gentes pluralitater
hebræos proprie noïamus aut ab Hebere ut ﹤
transitiuos significat.Soli qppe a creaturis n
nō scripta ad cognitioné ueri dei trāsiere:& u
ad rectam uitam puenisse scribunt:cum quil
totius generis origo Habraam numerādus es
iustitiā quā non a mosaica lege(septima eïm
Moyses nascitur)sed naturali fuit ratione cor
attestatur.Credidit enim Habraam deo & re
Quare multarum quoq; gentium patrem di
ipso benedicédas oés gentes hoc uidelic& ips
aperte prædictum est:cuius ille iustitiæ perfe
sed fide cōsecutus est:qui post multas dei ui
filium:quem primum omnium diuino psuar
cæteris qui ab eo nascerétur tradidit:uel ad r
eorum futuræ signum:uel ut hoc quasi pate
tinétes maiores suos imitari conaret:aut qbu

(5) *Detail from a page of Jenson's fifteenth-century types*

impressed into the matt surface of well-woven, rag paper is a tribute to past techniques in legibility – an affirmation of suitability. The printing from Caslon types was done in 1912 at the Chiswick Press. At the same time the printed pages of Nicolas Jenson (5) and Aldus Manutius (6) should be compared with this Caslon example, since it was these Italian models which influenced Caslon through Garamond and the Dutch types of van Dyck.

What are the factors in this kind of legibility? Sir Cyril Burt gave us a clue to personal preference based on familiarity – choice of type according to reading habits, implying maximum legibility in the most familiar letter-forms. Perhaps I should have re-read Sir Cyril – and the Medical Research Council's report on the legibility of print, Pyke's government paper of 1926. One reason why I have not done so is that these factors I want to identify and describe are not based on preferences for the familiar way in which mathematical or legal papers or the classics have been printed. The factors I wish to draw your attention to

fieri poſſe uix puto : ſed plane quia ita de-
bemus inter nos : neq; enim arbitror cario
rem fuiſſe ulli quenquam ;q̃ tu ſis mihi.
Sed de his et diximus aliâs ſatis multa ; et
ſaepe dicemus: nũc autem ; quoniam iam
quotidie ferè accidit poſtea, q̃ e Sicilia ego,
et tu reuerſi ſumus ; ut de Aetnae incendi-
is interrogaremus ab iis, quibus notum
eſt illa nos ſatis diligenter perſpexiſſe ; ut
ea tandem moleſtia careremus; placuit mi
hi eum ſermonem conſcribere' ; quem
cum Bernardo parente habui paucis poſt
diebus, q̃ rediiſſemus ; ad quem reiicien-
di eſſent ii, qui nos deinceps quippiam
de Aetna poſtularent. Itaq; confeci librũ;
quo uterq; noſtrum cõmuniter uteretur:
nã cum eſſemus in Noniano ; et pater ſe
(ut ſolebat) ante atrium in ripam Pluuici
contuliſſet; acceſſi ad eũ progreſſo iam in
meridianas horas die: ubi ea, quae locuti
ſum⁹ inter nos, ferè iſta ſũt. Tibi uero nũc
oratione utriuſq; noſtrũ, tanq̃ habeatur,

(6) *A page printed by Manutius of Griffo's*
fifteenth-century types

are basic to any language, any text, any age, and the critical moderator is the human eye. They are, and I dare to say it, instinctive, genetic, though perhaps also teachable, and implicit in the roman alphabet. All we need to know in order to make these factors self-apparent is how to use the alphabet.

So consider the transcription of a typescript into a sequence of printed pages – a process which should become increasingly simple as the designer's experience grows. Ideally, when shown a typescript, the designer should also have a *précis* of its contents so that, as he turns the pages looking for divisions and breaks, insertions and notes and quotations of one sort or another which may require special settings, he will have some idea of the construction and development of the text. He should also know, or be told, the age of the intended reader. With experience most of the work can be done without layouts.

In forming a clear idea of what the proposed book will look like, the first factor is the letter. By this I mean all that is implied by the choice of a typeface, or series of related faces which may include condensed or expanded letters, light or bold letters.

This choice involves a knowledge of the ratios of widths of main strokes to the height of letters, in addition to ratios of main stroke widths to crossbars, diagonals and thin strokes; stress of curves; types of serif; slant of italic and variation of slant; relation of x-height to extruders; presence or absence of irregularities of certain letters in the font; and the origin of the finally printed letters, that is to say, origin from metal types directly impressed, or repro-pulls photographed for lithographic plate-making, or from any of the various film matrix or CRT images.

The next three factors are:
the size of the letter,
the length of the line of letters,
the space between the lines of letters.

These are closely linked and should naturally be considered together. The size of the letter relates directly to the length of the line of letters (measure). The space between the lines of letters depends on the chosen letter, and the size of letter, and the measure.

The fifth factor relates to the space between the words – an important measurement closely inter-related with these first four factors. If spaces between words are not close, the eye reads slowly; if uneven, jerkily; but nevertheless the word-spacing has to be related to the character spacing, to the fit of the letters. The appearance of loose-fitting characters, as in the malpractice of letterspacing lowercase, reduces legibility, whilst tight-fitting characters may easily destroy the identity of the letters.

Next is the size of page, or format, which will take into account the length of the typescript, the kind of text that it is, and its use and market – respecting at least some of the conventions adopted by the printing and publishing trades.

The seventh factor, the printed area of the page, will largely be determined by the six preceding factors.

The eighth factor relates to margins surrounding the printed area on the page; this again will be affected by all decisions so far taken.

There is a subordinate consideration between factors seven and eight – belonging to both type-area and margins. I refer to extrusions beyond the regular printed area. These include folios, headlines, running headlines, shoulder titles, footnotes, catch-lines, and possibly antiwidow lines. It is convenient to consider at this stage half-titles and chapter headings and all forms of sub-headings, and we must not ignore the presence of signature marks and collating marks.

The ninth factor might be described as a visual or even mechancial aid to continuity of design and may vary from the simplicity of almost invisible pin-pricks in a mediaeval scribe's sheet of vellum, to a printed complex of guide-lines for text and picture areas in various relationships on the page – the grid. When designing a book the grid emerges naturally as requirements are put upon the designer by the nature of the text.

My tenth and last factor is 'finishing', already briefly mentioned, which includes how the work is imposed, printed and bound – giving details of paper, folding, collating, plating, sewing, trimming, blocking, covering, labelling, wrappering. Here marketing imposes itself because many buyers of the package are not the author's market, that is the reader, but intermediaries. They are the publishers' market, perhaps booksellers or librarians, or teachers, and in this process a more or less strong influence may be exerted by literary editors and their chosen reviewers.

The designer must work from the beginning with the finishing clearly in mind – all considerations, choices, decisions must support the total plan, but the plan is not inflexible; indeed it will often evolve as the answers are sought to the problems that arise along the way.

From the consideration of all these typographical factors will emerge a simple formula intended to answer the problems posed by the particular text in the interests of reader, author, editor, publisher, designer and printer. This formula, if successful, will provide legibility.

Whilst these details show something of the process of design, there is as yet no mention of style. In thinking about this I would like to propose a simple classification not by century or printer, not by divisions like symmetrical and asymmetrical, not by punchcutter or kind of letter, and not by its relationship to other arts of the period, but by analysis of the intentions of the designers. All typographical styles might be said to fall within five broad categories:

Full
Empty
Ornamental
Artificial
Natural.

Such a classification is valuable because it helps to simplify a complicated subject and may even influence some designers who show concern for the reader.

A series of title-pages will show what I mean. The title-page for William Morris' *Coleridge* (7) exemplifies *full* and is as full as any two-dimensional image can be. It leaves no doubt about stylistic intent. The opposite, or *empty* style, is well shown in a title-page (8) from the middle period of Tschichold. Then John Bell's Shakespeare of 1774 demonstrates *ornamental* (9) where the desire to decorate is more than evident. Of course ornamentation takes many different forms. *Artificial*, in this sense meaning contrived, is aptly shown in J. H. Mason's title-page (10) for Lawrence's *The Man Who Died*. Each line is made of uniform length irrespective of its relative value on the page by using a size of letter which will meet this demand – both word- and letter-spacing are grotesquely manipulated. Finally, the opposite of *artificial* emerges as *natural*, a typographical style in which the items on the page are arranged in a sequence of related editorial values which may be read vertically and where no item is too big or too small or overspaced or without adequate space. The University Press at Oxford has sometimes excelled in this style (11).

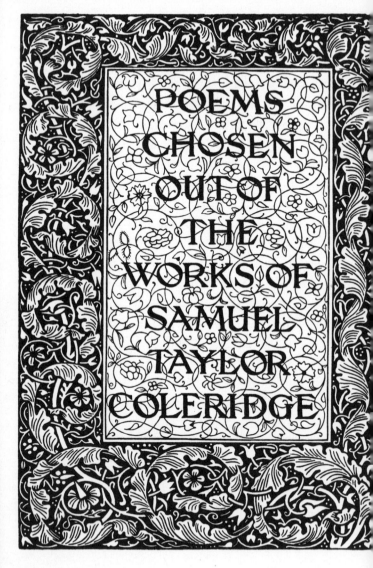

(7) *Style example* : FULL

Jan Tschichold

AN ILLUSTRATED HISTORY
OF WRITING AND LETTERING

LONDON · A. ZWEMMER

MCMXLVI

(8) *Style example :* EMPTY

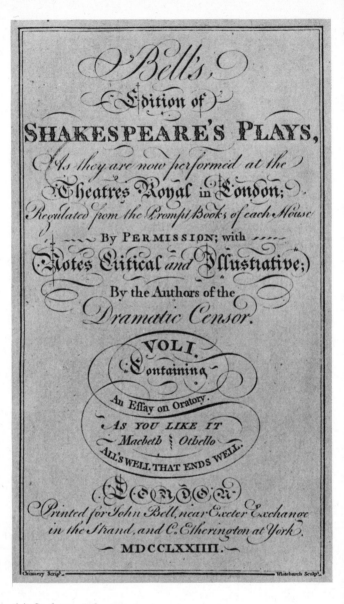

Bell's
Edition of
SHAKESPEARE'S PLAYS,
As they are now performed at the
Theatres Royal in London;
Regulated from the Prompt Books of each House
By PERMISSION; with
Notes Critical and Illustrative;
By the Authors of the
Dramatic Censor.

VOL I.
Containing
An Essay on Oratory.
As You Like It
Macbeth ⁂ Othello
All's Well that Ends Well.

London.
Printed for John Bell, near Exeter Exchange
in the Strand, and C. Etherington at York.
MDCCLXXIIII.

Grimmery Sculp.   Whitchurch Sculp.

(9) *Style example :* ORNAMENTAL

The Man who Died
By D. H. Lawrence

With illustrations drawn and engraved
on the wood by JOHN FARLEIGH

Type format arranged by J. H. MASON

Printed by W. LEWIS at Cambridge

London 1935 Published by William
HEINEMANN

(10) *Style example :* ARTIFICIAL

JAMES WARDROP

THE SCRIPT OF
HUMANISM

Some Aspects of
Humanistic Script
1460—1560

OXFORD
At the Clarendon Press
1963

---

(11) *Style example :* NATURAL

A semblance of style might be learnt easily enough, but I suspect that an understanding in depth is genetic, instinctive, and that natural style belongs to the organic make-up of a highly, visually, literate person, is as much a part of the natural order of sophisticated life as are the golden mean and other modules of proportional relationships. The discovery that masterpieces of the past relate to the golden mean, not accidentally but inevitably, is an important lesson for any designer, and it is particularly valuable for the typographer. Walter Kaech in his *Rhythm and Proportion in Lettering* gives analytical descriptions of, for instance, the Neptune temple (12) at

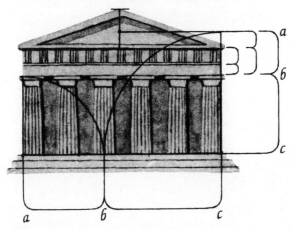

(12) *Golden mean analysis of the temple of Neptune, sixth century* BC

Paestum of the sixth century BC and of an inscribed tombstone (13) from Chieti of the first century AD.

The study of style is fascinating, and it stimulates creativity: but having glanced at it, we must get back to practical matters.

With design experience and a superficial knowledge of the typescript, the ten factors for legibility will have been weighed and a basic formula produced. The following example may be typical of well-made, general books:

Demy octavo, Monotype Series 169,
eleven on thirteen, by nineteen ems
by thirty-two lines plus folios
in spaced parentheses centred at foot.
Headlines centred in small caps spaced
two units.
Margins: inner, 4 ems; head, 5 ems.
Chapter drop, five lines.

Qualifying details and even layouts may be needed to complete the plan and to indicate what stylistic approach is being used, but the main structure of the book is established by that brief specification.

At least some of the many books on the setting and spacing and arrangement of type should be read; studying other people's ideas is a great help in working out our own. The exchanges of typographical ideas between Harold Curwen and J. H. Mason must at least be sampled by reading Leslie Owen's *J.H.Mason*, even though Leslie

(13) *Golden mean analysis of an inscription from Chieti, first century* AD

Owen's book is indulgent. B. H. Newdigate deserves some attention: the publication of his typographical principles in a literary magazine may well be unique. Joseph Thorp's *B. H. Newdigate* reprints many of these notes. The 1917 report on Cambridge typography by Bruce Rogers also merits a careful reading. It was made available to friends of the University Printer in 1950 and to the members of the Wynkyn de Worde Society in 1968. Eric Gill's *An Essay on Typography*, first issued in 1931, has points to compare with, for instance, Stanley Morison's *First Principles of Typography*, 1936, and with Oliver Simon's *Introduction to Typography* of 1945, and with Geoffrey Dowding's *Finer Points in the Spacing and Arrangement of Type*, 1954. Mardersteig, Rogers, Tschichold and Meynell have all published notes on the making of books which merit our attention.

In short, a comparative analysis of the way these designers have handled the ten factors I have enumerated, in theory as well as in practice, would be as valuable an exercise as one could devise for the aspiring typographer. Equally important in this process of laying the foundations of one's skills is an understanding of roots. Such an understanding is to be found in Harry Carter's *A View of Early Typography*, 1969. This introduction may give you a fresh insight into the period of incunable printing and certainly reveal the greatness of the Renaissance of printing at Paris in the 1520s. And if the past is important, there is the present to consider. An analytical approach to current work will also prove formative and will soon show that designer and editor must function neither separately nor in opposition but together if valid results are to be achieved.

This kind of analytical approach may be seen in a few examples of the contents pages (14) – (18) and at the same time the need to question certain traditional practices will be evident.

A second demonstration after the contents pages examines, for the same purpose, some details in the make-up and design of a small catalogue for an exhibition of William Coldstream's paintings (19) – (21).

North Wales Quarrying Museum,
London, 1975
Detail from a handbook shown here in
natural size. The original setting appears in
the top left-hand corner of a verso page,
208 mm deep. The appearance of page references where the eye expects to find chapter and section numbers is disconcerting,
and the main headings are weak and mis-
placed. The variable space between page
references and their items is also disconcerting to the reader.

Contents

Bloomsbury Portraits, London, 1977
Reduced from 245 mm deep. Constructed
on a formal, common, rigid plan which fails
to take into account the reader's need to
connect an item with its page reference
without the manual aid of a ruler.

Contents

(16) *Contents page*

Lines of the Alphabet, London, 1965
Reduced from 255 mm deep. Designed for
the reader, editorially and visually based on
the normal index arrangement for vertical
reading.

CONTENTS

(17) *Contents page*

Of the Decorative Illustration of Books Old and New, London, 1896

Reduced from 196 mm deep. Raises the following questions:

1. Is the heading prominent enough?
2. Is the heading swamped by the initial letter?
3. Is the repetition of 'chapter' required?
4. Are unspaced capital. letters sufficiently legible?
5. Do justified lines cause too great a variation in word spacing?
6. Without leading and with overspaced words the word space is sometimes at least three times the line spacing. A low legibility factor?
7. Is the folio below the decoration necessary?

CONTENTS.

(18) *Contents page*

Hugh Selwyn Mauberley, London, 1920
Reduced from 248 mm deep. This, an un-
happy-looking page, has at least a score of
typographical blemishes. It needed a firm
editorial hand to present the material as an
index of first lines (which it could have
done) rather than as a list of contents (which
it has failed to do).

MAUBERLEY

CONTENTS
Part I.

ENVOI
1919

Part II.
1920
(Mauberley)

(19) *Catalogue cover with comment*

(20) *Catalogue title-page*

(19) – (21) *Catalogue*

By accident and almost unwillingly I made notes on this catalogue at the exhibition of paintings by William Coldstream. So many were my differences with the designer of the catalogue that I decided to record them for teaching purposes. On the front cover (19), reduced from 190 mm deep, comment is drawn.

On the title-page (20) head and tail margins fail to contain the words on the paper in anything like a satisfactory *mise en page*. In detail, the omission of punctuation is not in keeping with the text. The postcode in the Scottish address overshadows the word 'Edinburgh' but no code is given for the London address.

The centre opening (21) reveals large ugly staples and at first sight appears to be a wrongly folded sheet. But if one mentally adjusted the text areas into a better position the folio on page 6 would be trimmed off. It is simply that a grid of unhelpful proportions has been adopted.

sitter, to a large extent, dictates the pose, and you have to start painting before very long. But, if you're going to go to your studio and say 'What am I going to do?', there seem so many alternatives that one's much slower getting started.

"I've always found the sitters I've had have been extremely considerate. Mind you, I do warn them beforehand that I want a lot of sittings, so that they know what they're in for. But I have found that they have been pretty considerate, and they've been willing to go on sitting for a great many times. I don't like showing them the picture while I'm doing it, and they nearly always are extremely tactful about not asking to see it. Once I've shown a picture to the sitter, I find it very difficult to go on, because, supposing they like it, or at any rate don't dislike it too much, or say that something's nice about it, then you feel that when you've gone on they'll look at it and probably think it's not so nice, so that it somehow inhibits one. And I think that, while one's painting on a picture, it always seems to me that every time you start you've got to be willing to risk everything, and if there's a passage you happen to like, you always feel you've got to be prepared to put your brush right through it. You never have to have any regard for the past when you're painting. I think you've got to start work every time as though you were starting absolutely afresh, and risk everything. And the moment you feel, 'Well, I don't want to spoil this', you are well on the way to spoiling it. And, once you've shown it to a person, if they say there's something wrong with the mouth or the eyes or something, it's impossible to alter it, because I don't paint like that. I mean, the mouths and eyes, as it were, accrue – they're not put there, in my way of painting. And therefore I can't really do anything about the expression, as they say. So, even if I wish to, I can't really alter it to their requirements, and if they like it at all, or seem to, then I am rather afraid of spoiling it."

The process he describes here – and most of it still applies when the model is being paid by him – is an interaction between rigorous application of the artist's will – the insistence on numerous sittings, the frustrating of the sitter's curiosity, the readiness to risk spoiling what is there – and a passive, stoical acceptance of what happens beyond his control – the model whose appearance defines the task, whose arrival compels an immediate start, whose comfort decides the pose; above

all, that things aren't 'put there' by him, that they "accrue". And they must at all costs accrue. So long as they do accrue, he is painting forms, not features, painting what he sees, not what he knows. And Coldstream employs a ritual which must help towards painting mindlessly, so to speak – the practice of 'measurement' by which the painter holds out a brush or a plumb-line at arm's length and marks out intervals with one eye shut ('as if you were shooting'). In its sustained watchfulness and in its objectivity, measurement is the antithesis of techniques which introduce the workings of chance into the operation (whether active methods, such as Bacon's throwing of paint, or passive ones, such as Duchamp's leaving the surface to accumulate dust), but, by interposing a mechanical process between model and artist, serves the same purpose of ensuring that the ego does not take control but enters into a collaboration with phenomena that have an impetus of their own.

"I get intense pleasure when I'm painting in just saying that this is two-thirds that, that this is one-sixteenth more than that. Now, if you get a number of measurements, both on the surface of a canvas and then imagined measurements to some extent in depth working together, it gives you a kind of kick – I don't know why, that then this system builds up and you build on to it. I can't rationalise it. But it's a kind of playacting it seems, in a way, which gives one pleasure, somehow seems important. I'm always driven back to it because it seems to me to be extremely interesting to ascertain something. I know, of course, this is childish, because the measurement in a sense is conventional. But the idea of making sure about the thing, within the limits of the game, means a lot to one.

"When I'm starting a painting, I paint quite a long time freely without measuring now. But I'm driven back to measuring. All this process, I suppose, in a way is to get something on the canvas which you can believe in – to get some knot of life or something going which will have some power of generating itself, you see. It seems less willed if you're measuring it. If you're trying to get something going by your own will power, you feel you've got to, refer, to something which seems impersonal, it gives me a sense of distaste. If you have this rather sort of arbitrary thing, you feel the problem that's coming out is not one which you've been too much consciously associated with and therefore has that attraction."

As to measuring in order "to get something on the canvas which you

(21) *Centre opening from the Coldstream catalogue*

The measure is too wide for unleaded lines and why are the lines unspaced when, in this ten-page booklet, there are more than four pages of white paper? And why is art paper used when there are no illustrations? Extra spaces appear between certain paragraphs yet, within the six pages of text, new paragraphs begin on tail-lines. The title-page, the text page and the list of pictures use different grids and all are unsatisfactory.

I have several catalogues from the gallery of d'Offay Couper (more recently Anthony d'Offay). Strangely, the only one in my collection to have numbered pages is the *Coldstream* catalogue. The *Claud Lovat Fraser* catalogue of 1971 has 56 pages without folios. Nevertheless, the exhibitions and their catalogues are usually of special interest and quality. For this reason the *Coldstream* item is noticeable.

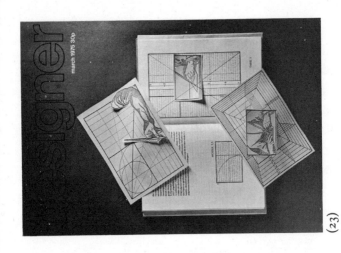

(23)

(22)

(22) *Analytical grid from Rosarivo's 'Divine Proportions in Typography'*

(23) *Cover for 'The Designer' based on Rosarivo's book*

It is all too common for the reader to experience grids of unhelpful design, grids that force the beginnings or ends of textlines into the gutter margins allowing them only reduced visibility.

Or a grid may be brought to bear analytically upon an incunable in much the same way as the golden mean might be used to show that certain proportions have always existed in Nature. The incunable shown here (22) is Fust and Schöffer's *Psalter* of 1457 and the grid is from Rosarivo's *Divine Proportions in Typography*. A grid might even be used decoratively as a symbol with other graphic symbols. A cover for *The Designer* (23) was a by-product in the making of a symbol for articles in *The Times Literary Supplement*.

But of course the grid is normally devised to aid the planning and make-up of pages. We use it as a control; we must not let it control us. The exercise of control is for the reader's sake – for the sake of legibility.

Perhaps the most important grids form themselves out of experience, preference and function and are carried in the head.

They are, as it were, made from sliding, overlapping, transparent and infinitely adjustable modules and means.

Of course the typographer has first to develop a style (which itself must be adaptable) before he can use such mental resources. Then he is fully equipped to design pages of a book out of sequence and yet make every part of it fit together in unity.

I don't wish to make the act of designing sound like a metaphysical exercise. Rather it is like a process that comes naturally, purposefully, as a result of putting oneself in the place of the reader. With this approach a workable draft can be easily, simply sketched (24). This rough plan was drawn for the Bodley Head Christmas booklet for 1976.

Directing the production of bookjackets is a special challenge, calling for an ability to combine explicit pictorial images with legible drawn or typographical lettering. For instance, from a typographical reconstruction (25) of a Roman tablet by Johann

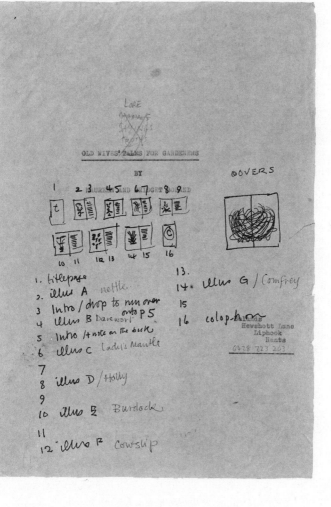

LORE

~~garrias~~
~~for~~
~~boy~~

OLD WIVES' TALES FOR GARDENERS

BY

1   2   3   4   5   6   7   8   9

COVERS

10 11   12 13   14 15   16

1. titlepage
2. illus A   nettle.
3. Intro / drop to run over
4. illus B Darewort   onto P 5
5. Intro / note on the back
6. illus C Lady's Mantle

7.
8. illus D / Holly
9.
10. illus E Burdock

11.
12. illus F Cowslip

13.
14. illus G / Comfrey
15
16 colophon
                Hewshott Lane
                Liphook
                Hants
                0428 723 203

(24) *First sketch for The Bodley Head Christmas booklet for 1976*

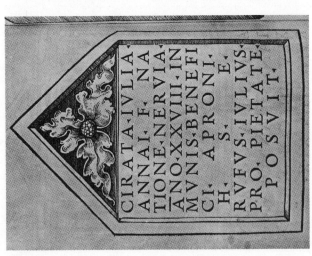

(25)

(26)

(25) *Typographical reconstruction of a Roman tablet of Johann Schöffer*

(26) *Book jacket by Michael Harvey based on Schöffer*

Schöffer which I found in Harry Carter's *A View of Early Typography*, I had no problems in directing Harvey's transcription of it into a jacket (26) for *The Idea of the City in Roman Thought*.

The jacket for *An Impossible Woman* began with a painting of the author being serenaded but under pressure from the sales department the idea was abandoned in favour of a lettered design which went through a sequence of changes, three of which are shown here (27) – (29).

(27) *Michael Harvey's first sketch*

(29)

(28)

(28) *Harvey's second rough*

(29) *The final arrangement of lettering for 'An Impossible Woman' jacket by Michael Harvey*

When a book remains in print for many years – *Ulysses*, for instance – the jacket may pass through a number of transformations. From Paris in 1922 *Ulysses* appeared in blue paper covers (30) with reversed typographical lettering on the front and no lettering on the spine or back. The same covers were

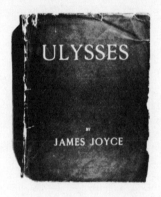

(30) '*Ulysses*' *first edition cover, 1922*

used for John Rodker's edition. From Hamburg in 1932 the two-volume paper-covered edition had drawn lettering on front and spine printed in red-brown ink on grey paper. The first Bodley Head edition in 1936 was wrappered in plain paper with title, author and Eric Gill's long-bow printed in black.

(31) *Michael Harvey's sketch for the 1977 impression*

Later editions, cut down in size and quality by the war, had green paper jackets printed on front and spine from black type, and after the war The Bodley Head, under Max Reinhardt's direction, re-set the text in a new format with a similar jacket but with Gill's long-bow in white on the spine. Then *Ulysses*, for a limited period, became the-book-of-the-film and appeared in disguise with details from stills on the front and back. The first Penguin edition was issued in 1968 with lettering reversed white out of a black cover. The wrapper for the latest impression from The Bodley Head is by Michael Harvey (31) with the base colour in dark blue and the lettering in white and pale yellow.

A design that can hold its own identity and message and be instantly recognisable in miniature is probably good. The Stephen Russ jacket for Graham Greene's *May We Borrow Your Husband?* certainly is. The detail from an advertisement for cotton is reproduced in natural size (32) and the final drawing by Stephen Russ is reduced from 203 mm high (33).

(32) *From an advertisement for cotton featuring the jacket for 'May We Borrow Your Husband?'*

(33) *Stephen Russ's drawing for the jacket featured in* (32)

Sometimes the making of a jacket becomes an involvement between author and publisher where art directives are simply trying out, or trying to follow, a set of proposals. The first suggestion came from Graham Greene for his novel, *The Honorary Consul*, and this was to use a drawing of maté (34) since the plant could be a clue to

(34) *Stephen Russ's drawing of maté for 'The Honorary Consul'*

(36)

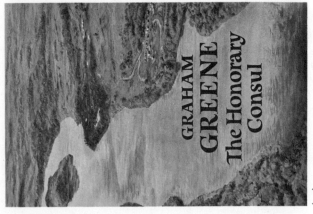

(35)

(35) *A drawing of the Parana River for 'The Honorary Consul' made from a colour film*

(36) *Norstedts' design for 'The Honorary Consul'*

(37) *Michael Harvey's lettering for 'The Honorary Consul'*

the location of the novel. Then there appeared in one of the Sunday colour magazines a photograph (35) of the location on the Parana River, an even more suitable location clue. Meanwhile a proof of the Scandinavian lettered jacket (36) appeared in the publishers' office. And when the transparency of the river scene was examined it had to be discarded. A bridge on the photograph contradicted the plot of the novel. So a drawing was made from the transparency without the bridge but this too was shelved in favour of a lettered jacket (37) by Michael Harvey.

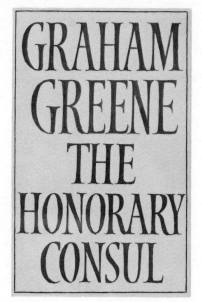

(37)

for colours please see
separate colour swatch

(38)

The wrapper for *Saint Jack* had no such complications. Only one rough was drawn and the triple image then masked to a girl's head and shoulders (38). To the final

drawing I added the lettering (39). Usually
Stephen Russ includes lettering as part of
the jacket design as on the one for *Rage*.
Shown here (40) is his rough drawing.

(39) *Stephen Russ's jacket for 'Saint Jack'*

(40) *Stephen Russ's final rough for 'Rage'*

(40)

Long-standing and good relations between the jacket designer and the publisher's
designer help to solve jacket problems.

PRINTING
AND
DEMOCRACY

BY ROBERT BIRLEY

C R

LONDON
PRIVATELY PRINTED FOR
THE MONOTYPE CORPORATION LTD
1964

(42)

HIS MAJESTIES

DECLARATION

To all His

Loving Subjects, A

*After his late Victory against the Rebels,
on Sunday the 23. of October.*

Together with

A RELATION OF

The battell lately fought betweene *Keynton*
and *Edge-hill*, by His Majesties Armie, and that of the Rebels.

*With other successes of His Majesties
Armie happening since.*

Charles R.

OUr expresse pleasure is, That this our Declaration be published
in all Churches and Chappels within the Kingdome of *Eng-
land* and Dominion of *Wales*, by the Parsons, Vicars, or Curates of
the same.

Printed by His Majesties Command at *Oxford*,
by *Leonard Litchfield*, Printer to the *Universitie*, 1641.

(41)

(41) *Title-page printed at Oxford in 1641*

(42) *Booklet cover printed at Oxford in 1964*

However, some textual legibility problems remain, and an attempt to establish a touchstone for those of us involved in perpetuating the written word may be helpful.

Myfanwy Piper once wrote of Reynolds Stone: '. . . he had a responsibility towards each letter each time he cut it so that its full dignity and character was brought out.' Such responsibility can hardly be claimed for dry-transfer lettering manufacturers who promote student lettering of maximum illegibility.

Between these two extremes we have to maintain visual excellence in the wake of computer-organised filmsetting which includes font development or the making of new alphabets by electronic methods.

It is important for us to pursue legibility rather than be forced to join a society for the prevention of cruelty to human eyesight. We must see to it that our editors re-arrange copy which might result in confusion (41) so that it can be presented ready for reading vertically (42) like the cover to a booklet printed at the Oxford University Press. Also we must encourage designers to employ their freedom in the pursuit of legibility. Printing with capital letters can be done sufficiently well to arouse interest (43), (44) and, with short lines, reading at a slowed speed is possible – but in principle too many factors of low legibility are involved.

When we think of the triumphant collection of twenty-six unaccented roman letters – the simplest known key to literacy? – we realise that there is an overwhelming advantage inherent in the roman alphabet and particularly for us in the English language. We have miraculously avoided the oriental policy that has forced millions of readers to live in a maze of pictographs. We have even avoided the legacy of Gutenberg's 290 separately cast characters. So why do we join letters, fuse them together and add unwise spaces that can only distress the reader's eye? We even print words in unacceptable colours reversed out of screaming backgrounds. Instead of straining for effect in such ways, we should be examining every detail of legibility with the discrimination that results from understanding the roman alphabet.

OF THE TRUE GREATNESSE OF KINGDOMES AND ESTATES

THE SPEECH OF THEMISTOCLES THE ATHENIAN, WHICH WAS HAUGHTIE AND ARROGANT, IN TAKING SO MUCH TO HIMSELFE, HAD BEEN A GRAVE AND WISE OBSERVATION & CENSURE, APPLIED AT LARGE TO OTHERS. DESIRED AT A FEAST TO TOUCH A LUTE, HE SAID; HE COULD NOT FIDDLE, BUT YET HE COULD MAKE A SMALL TOWNE, A GREAT CITTY THESE WORDS (HOLPEN A LITTLE WITH A METAPHORE) MAY EXPRESSE TWO DIFFERING ABILITIES, IN THOSE THAT DEALE IN BUSINESSE OF ESTATE. FOR IF A TRUE SURVEY BE TAKEN, OF COUNSELLOURS AND STATESMEN, THERE MAY BE FOUND (THOUGH RARELY) THOSE, WHICH CAN MAKE A SMALL STATE GREAT, AND YET CANNOT FIDDLE: AS ON THE OTHER SIDE, THERE WILL BE FOUND A GREAT MANY, THAT CAN FIDDLE VERY CUNNINGLY, BUT YET ARE SO FARRE FROM BEING ABLE, TO MAKE A SMALL STATE GREAT, AS THEIR GIFT LIETH THE OTHER WAY; TO BRING A GREAT & FLOURISHING ESTATE TO RUINE AND DECAY. AND CERTAINLY, THOSE DEGENERATE ARTS AND SHIFTS, WHEREBY

76

(43) *Text-page printed at The Shakespeare Head Press in black and red for the Cresset Press, 1927*

C A P V T   I I I.

Pleriqve eorvm, qvi veteris poe-
ticae historiae fastos non satis
accvrate dispicivnt, in evm dela-
bvntvr errorem, vt credant, at-
qve affirment, anacreonta mise-
re arsisse sapphvs amore. sed pv-
gnat aperte ratio temporvm, qvi-
bvs vixisse tradvntvr. nam, qvem-
admodvm iam monvi, vates vi-
tam agebat svb polycrate, et cy-
ro, lesbia vero poetria, qvae so-
ror addita mvsis dicebatvr, vi-
xit svb alyate craesi filio. nec
vllvm peti poterit argvmentvm
ad erroris excvsationem ex testi-

(44) *A page printed by G. B. Bodoni at Parma in 1784*

(45) *Detail from a page printed by Castellione in 1541*

I have tried to suggest that it is time to
pause, to give some fresh thought to the
major changes that have occurred in typography.

I am hinting at the addition of sloped
roman capitals to the font of italic, and this,
I think, was a mistake. The use of small up-
right roman capitals with italic lowercase
provided the perfect amount of irregularity
the eye needs for maximum legibility. Add
to this, minimal changes in slant and re-
strained flourishes as we see in a page (45)
printed by Castellione of Milan in 1541 and
you have all the contrast needed when used
with a text set in roman letters.

Humaniſsime Lector Bonauenturam Caſtil
in Templo Scalæ Mediolani Canonicum;
·um Regionem tot Sæculis ab omnibus ferè
1 Gręcis tum maximè Latinis Silentio fer-
n veluti è Ténebris nunc ereptam;in Lu-
Alpes et colles ad Inſubriám Spectantes
totam planitiem quę inter Ticinum et Ab-
d padum- vſq₃ cõtinetur; vrbem demum
m olim Symetriam referentem treis in li-
,Adiecta Inſuper Tabella quam mappam
1 omnia hæc continentur. Gratulabar igi-
(45)

Variation of slant is a vital factor in legibility which must not be confined to the lowercase f – the habitual casting convenience on the Linotype. A solitary outbreak of this sort is eccentric and can only be objectionable to the eye. What is needed is an organically systematic variation of slant in italic lowercase. In other words, we must avoid such typographical mistakes as resulted from the recutting of Griffo's 1501 italics in which all mainstrokes were made exactly parallel. A similar loss of identity and legibility occurred in the recutting of Anton Janson's italics. The three illustrations (46) – (48) show how one version of Janson with its several angles of slant has

The attention of the reader is drawn to the fact that in the facsimile the calligraphic specimens are numbered in pencil by hand to allow the reader to follow the references made in the text to individual plates.

(46) *Janson italic printed by Giovanni Mardersteig*

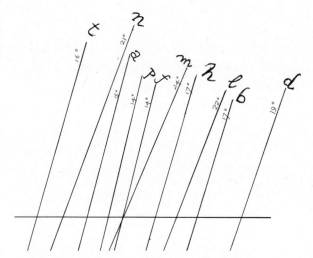

(47) *Slant analysis of the Janson italic shown opposite*

been mis-cut with a single, constant nineteen degrees slope. Both these recuttings of italics, Bembo and Janson, suffer from the lack of variation of slant. As a result they have become less legible.

abcdefghijklmn opqrstuvwxyz

(48) *Janson italic with uniform slant of nineteen degrees*

Let us introduce Nicolas Jenson of Venice and Francesco Griffo of Bologna and the Parisian Simon de Colines with his young successor Robert Estienne to the computers before it's too late. Let us programme with the few golden canons of typography we believe in.

COLOPHON

The Case for Legibility is set in Monotype Ehrhardt
metal letters, repro-proofed and printed together
with the illustrations by offset lithography at the
Stellar Press. The text paper, Vintage Esparto Cartridge, is made by East Lancashire Paper Mill and
supplied by A. H. James & Co. Ltd. The book has
been bound by Leighton Straker Bookbinding Co.
Ltd in Dover cloth supplied by
Grange Fibre Co. Ltd.